Charles Smith Boynton

Chemical Questions and Problems

Charles Smith Boynton

Chemical Questions and Problems

ISBN/EAN: 9783742817723

Manufactured in Europe, USA, Canada, Australia, Japa

Cover: Foto ©Angelika Wolter / pixelio.de

Manufactured and distributed by brebook publishing software
(www.brebook.com)

Charles Smith Boynton

Chemical Questions and Problems

Chemical

Questions and Problems

By C. SMITH BOYNTON, A. M., M. D.

Instructor in Chemistry,

Medical Department, University of Vermont.

BURLINGTON, VERMONT.

1891.

General Properties of Matter.

What is Chemistry?

How may natural phenomena be divided?

Explain the difference between a physical and a chemical change?

What experiments prove this?

By what agents may chemical action be determined?

What experiment shows that chemical action has taken place through the action of mechanical force?

What experiment shows that chemical action acts only at immeasurably minute distances?

Describe the experiments with S and Fe, showing the difference between a physical and a chemical change; a mechanical mixture and a chemical compound?

What name is given to this chemical compound?

What is the difference between a mixture of 12 parts by weight of powdered charcoal with 64 parts S and the liquid known as carbon disulphid?

By what are chemical changes almost universally affected?

Of what is the sensible universe made up?

What is matter?

What is meant by the term Indestructibility?

What by Impentrability?

How many forms of attraction in science?

Give examples of each?

What is meant by homogeneous, and hetrogeneous matter?

How many kinds of weight are recognized in chemical operations?

Explain absolute and apparent weight?

Which system of weights and measures is employed in chemistry and chemical operations?

Give the subdivisions of a metre?

What are the decimal multiples of the metre?

What is the unit of fluid measure in the metric system?

How many ounces in one litre?

Give the unit of metric weight.
> (One gram, equal to 15.432 grains.)

How is the gram derived?
> (From a cubic centimetre of distilled water at its maximum density $+ 4^\circ$ C.)

What are the decimal subdivisions of a gram?
> (1 gram = 10 decigrams = 100 centigrams = 1000 milligrams.)

How are the decimal multiples of a gram designated.
> (By Greek numerals. 10 grams = 1 dekagram; 100 grams = 1 hektogram; 1000 grams = 1 kilogram.)

Give the equivalents in Troy weight of 1 gram, 1 kilogram, also in grams of 1 grain and 1 oz. Troy.
> (1 gram = 15.434 grains; 1 kilogram = 2.679 lbs. Troy; 1 grain = 0.065 grams; 1 oz. Troy = 31.103 grams.)

Describe the chemical balance?

4

How should chemical balances be kept?

By what means are minute differences of weight determined?

When a body is weighed in air, and then in water, where is it the heaviest?

What is the loss of weight it experiences in water?

What is specific gravity or specific weight?

What is the standard for Solids, Liquids, and Gases?

The specific gravity of sulphuric acid is 1.8; what is meant by this statement?

What is the standard temperature adopted by the U. S. P. for the determination of specific gravity?

What is the temperature of water at its greatest density?

What is the object in determining the specific gravity of urine?

What is meant by the expression that one substance is lighter than another?

State the difference between density and specific gravity?

What is the weight of one cubic foot of platinum, at 60 ° F.?

Give the weight of a cubic foot of turpentine, at 60 ° F., and state how the result is obtained?

What is the standard barometric pressure?

Describe the specific gravity bottle, or Pykno-meter?

If a bottle, capable of holding 300 grains of distilled water, holds 260 grains of some other liquid, what is the sp. gr. of that liquid?

How may the weight of any substance be calculated if the volume and specific gravity be known?

Required the weight in pounds of a dry block of fir containing 50 cubic feet, the sp. gr. of the wood being 0.47?

How is the specific gravity of a body heavier than water determined?

How is the specific gravity of solids in the form of powder or fragments determined?

If 100 grains of a solid be introduced into a bottle holding 500 grains of water, and after the introduction of the solid the bottle weighs 560 grains, what is the sp. gr. of the solid?

If a body is soluble in water how is its specific gravity obtained?

If 200 grains of a solid be introduced into a bottle holding 400 grains of alcohol, the sp. gr. of which is 0.870, and the bottle weighs after the introduction of the solid 570 grains, what is the sp. gr. of the solid, taking water as the standard of comparison?

If a body is lighter than water, how is its specific gravity obtained?

How is the specific gravity of gases obtained?

In how many forms does matter exist?

On what does the different forms of matter depend?

To what are the repulsive forces due?

Define divisibility and give examples?

Physical Characters of Chemical Interest.

What is meant by crystallization?

What is an essential condition of crystallization?

What is the axis of a crystal?

How may a crystalline form be defined?

What is meant by cleavage?

When are bodies said to be amorphous?

Into how many systems may the different crystalline forms be classified?

State the different systems, and their characteristics?

What is meant by Isomorphism.

Give examples of isomorphous minerals.

Define dimorphism.

What is meant by water of crystallization?

What is meant by the effloresence of a crystal?

Define allotropism?

On what may differences of the allotropic state depend?

What is the difference between solution and chemical change?

To what substances is the term colloid applied?

What is a crystalloid?

What is dialysis?

Define the meaning of diffusate and dialysate?

•Describe the dialyser?

What is latent heat?

How much heat is generated by the union of one pound of H with eight pounds of O to form water.

Explain the principle of the so-called freezing mixtures?

Suppose a definite amount of heat to be imparted to a mass of lead, how is that heat disposed of within the substance?

Describe the experiment illustrating the differences which exist between bodies, as to the quantity of heat they are able to take up?

What is specific heat or " capacity for heat"?

What relation exists between specific heat and atomic weight?

How do we measure the temperature of bodies?

Explain the difference between a thermometer and a pyrometer?

What is the best thermoscopic agent?

When are alcohol thermometers to be preferred?

What are the principal scales of the thermometer in use?

How is the Fahrenheit scale graduated?

How is the Centigrade scale graduated?

Which is commonly used, and which in chemistry?

How are degrees of Centigrade scale converted into those of Fahrenheit?

How are degrees Fahrenheit converted into Centigrade?

What results when a beam of light passes through a glass prism, and falls on a white screen?

What effect has the prism on the sunbeam?

If instead of sunlight, we use some colored flame with the prism what is the result?

What is spectrum analysis?

Describe the spectroscope?

What is an absorption spectrum?

What is a continuous spectrum?

Describe the experiment proving that "substances when cold absorb the same rays which they give out when hot"?

What conclusion is reached from a study of the sun's spectrum?

Also of the several star spectra?

What is polarized light, and how does it differ from ordinary light?

What is meant by refraction?

Define the terms, polarizer and analyzer?

What is meant when a substance is described as optically active?

Explain the terms dextrogyrous and loevogyrous?

What is meant by the specific rotary power of a substance?

State examples of the chemical effects of light?

What rays of the solar spectrum possess the greatest chemical activity?

How much of the spectrum rays are invisible as light?

How are these rays recognized?

Describe the construction of a galvanic battery?

What effect has the "current" on a magnetic compass?

Explain the terms, negative pole, positive pole, cathode, electrodes?

What comprises a galvanic circuit?

What is meant by an open and a closed circuit?

Define electrolysis?

What is an electrolyte?

What are the products of the decomposition termed?

Chemical Combination.

How may matter be subdivided?

Define simple and compound matter?

How many elements are at present known?

How many of these elementary substances possess metallic properties?

How many, and what elements are gases under ordinary conditions?

How many and what elements are liquids?

Which are the lightest, and which the heaviest?

How are these elements distributed in nature?

How many of these elements make up ninety-nine one-hundredths of the whole mass of the earth?

What is meant by analysis?

What by synthesis?

What is meant by qualitative and quantitative analysis?

What is the law of definite proportions?

What is a chemical compound?

What is a mixture?

What aid is given by the law of definite proportions, for distinguishing between a mixture and a true chemical compound?

State the law of multiple proportions?

How is this law illustrated by the compounds of N and O?

State the law of reciprocal proportions?

How is matter divided?

What is a mass of matter?

What is a molecule?

What is an atom?

How should the word atom only be used?

To what does the term molecule apply?

What is meant by atomic weights?

What is the difference between the molecules of an elementary substance and those of a compound?

State the law of Gay Lussac?

What is the kinetic theory of gases?

State the law of Avogadro?

How can this law be proven?

To what is the weight of any molecule compared with that of hydrogen proportional?

Give an example?

Explain why the molecular weight of H is 2?

Why do molecules differ from each other?

How may atoms differ from each other?

Give examples of differences of atoms in kind, number, and relative position?

What is the atomic weight of an element?

State the observations of Dulong and Petit?

What is atomic heat?

What results from the study of atomic heats as shown in the elements Bo, C, Si, S, and P?

Define molecular weight?

How is the molecular weight of a substance obtained?

Explain the system of chemical notation?

In writing the symbol H for hydrogen, what does it represent besides hydrogen gas in general?

In the same way, for what do the symbols O, N, C, stand respectively?

Explain HNO_3?

How are chemical formulæ written?

Write the formula showing Zinc acted upon by sulphuric acid?

Also showing Potassium nitrate acted upon by sulphuric acid?

Complete the equations $Zn + H_2SO_4 =$ and $KNO_3 + H_2SO_4 = ?$

What is meant by stoichiometry?

What fundamental principle is to be remembered in writing formulæ?

What is meant by valency?

What element is used as the standard of comparison?

On what do the proportions in which atoms combine depend?

What is the meaning of univalent, bivalent, trivalent, quadrivalent?

What names are given to the atoms with reference to their valency?

Some metals do not combine with H, how is their valency determined?

How is the valence of an element represented?

What is an unsaturated compound?

Give an example of a triad united with a monad and a dyad in such a way that the valancies exactly balance?

13

How are the molecules of free H_2O, and N to be regarded?

What is meant by the function of a substance?

What is the effect of Na on water?

What are the products of the reaction?

How is this proven?

Express the products of the reaction by chemical symbols?

How can you prove that the formula is not $Na_2 O_2 H_2$?

What is meant by metathesis?

When we say that this reaction between the atoms of Na, each with an atom of H, is a simple example of metathesis, what is meant?

What is meant by the term hydrate?

Give the formulæ of Potassium hydrate, and Sodium hydrate, and show that they have the same type of structure as water?

What other name is given to the hydrates?

What is the product of $NH_3 + H_2O$?

What is there remarkable about this compound?

Which of the hydrates have important applications in the arts?

How do they differ from the hydrates of the other metals?

What class of compounds possess characteristics opposed to the alkalies?

What features have these substances in common, which lead chemists to call them all acids?

What is the effect of adding the acid to the alkaline solution?

When the sodium solution is treated with the acid HCl until it has no effect on the test paper, what is the product obtained on evaporation?

State this reaction in the form of an equation?

What is done when an acid is neutralized by an alkali?

If a solution of Potassium hydrate is neutralized by nitric acid, what is produced on evaporation of the solution?

What is the chemical formula for nitric acid?

State the molecular weights of H, N. and O?

Give the molecular weight of nitric acid?

What two principles can be deduced from these experiments. as true in the case of all alkalies, and all acids?

What then is an acid?

Some chemists regard an acid as a "hydrogen salt," from this point of view how would you name sulphuric, and nitric acids?

Upon what does the characteristic qualities of an acid depend?

Is this shown in any other way than the special mode we have been studying?

Give examples of experiments showing this?

In the experiment with ZnO and H_2SO_4 why is there no H evolved?

State the equation showing the reaction?

What was the idea of Lavoisier regarding O and acids?

15

Was he right?

Are there any acids that contain no O?

What terms have been applied to the oxygen and hydrogen acids?

What is meant by the basicity of an acid?

Give the meaning of the terms monobasic, dibasic, tribasic, and polybasic?

What is a base?

What is a double decomposition?

What are sulphobases or hydrosulphids?

What facts are to be learned from the study of the act of neutralization?

What are the products of the action of an acid on a base?

What is a salt?

State the difference between the haloid salts and the oxy salts?

Define a neutral salt, a basic salt, and an acid salt, giving examples and formulæ of each?

What is a sulphur salt?

What relation do these salts bear to the corresponding oxygen compounds?

Give formulæ and examples?

Define an anhydrid?

Name some of the most important?

What is produced when baric oxid burns in the vapor of sulphuric anhydrid?

What does this show?

What is meant by the term metal as used in chemistry?

What is a radical?

As a general rule, what distinction may be observed?

Of the two radicals containing N (NH_4 and CN) which are basic and which acid?

On what do the valances of the radicals depend?

How are the radicals usually written in an equation?

How do the names of radicals terminate?

Are the terms radical and residue synonymous?

Give an example to prove the correctness of the last answer?

What is the principle of the present system of nomenclature?

How are binary compounds indicated?

For what terminations has the present *id* been substituted?

Give examples?

Why is the older name retained in some instances?

Give examples?

When two elements unite with each other to form more than one compound, how are these distinguished from each other?

What other method is used for distinguishing two compounds of the same elements?

Give examples?

How are the oxacids named?

Explain the use of the suffix *ous* and *ic*?

If there be more than two acids, formed in regular series, how are they named?

Read the formulæ $HClO$, $HClO_3$, $HClO_4$, $HClO_2$?

How are the names of the oxysalts derived?

How are acids whose molecules contain more than one atom of replaceable hydrogen distinguished?

Give examples?

Certain metallic elements form two distinct series of salts, how are these distinguished?

To what has the names, basic salts, subsalts, and oxy salts been applied?

What is meant by double salts?

Give an example?

What is meant by the composition of a compound?

What is the constitution of a compound?

What theory is now advanced regarding the motion of the atoms in the molecule?

In the classification of the elements, how are they divided?

Name the primary divisions of the four classes?

How are these classes divided?

What are contained in each group?

Hydrogen.

Where and in what form is H met with in nature?

How does it occur chiefly on the earth?

How is H obtained?

What takes place when a piece of sodium is thrown upon water?

What takes place when steam is passed over heated iron?

Give the equation representing this reaction?

What is theoretically the simplest method to prepare H?

What is water gas, and how is it obtained?

What is the most convenient method for preparing H?

Describe the process fully, and show the method of collecting the H?

What are the properties of H?

What are the common acids?

What do they all contain, what takes place when they are treated with an element?

What relation is there between the weights of equal volumes of H and O?

If the weight of a certain bulk of H is one ounce, what would be the weight of the same bulk of oxygen?

How is it shown that H is lighter than the air?

19

What is meant by the diffusion of gases?

What relation is there between the specific gravity of gases and the rate at which they diffuse?

What is the law governing these phenomena?

By this rule, how much faster will H diffuse than O?

Does H combine with O at the ordinary temperature?

Does H support combustion?

How can this be shown?

What is reduction?

For what purpose is H often used in the laboratory?

What is a reducing agent?

What is an oxidizing agent?

Is H ever absorbed by the metals?

Give examples, and state under what conditions it takes place?

How many times its own volume of H will palladium absorb under favorable conditions?

What changes are produced in the metal?

What name has been given to this absorbed form of H?

What is the weight of one litre of H?

What is this weight called?

What is the law governing the expansion of gases by heat?

How much will a gas expand for each degree centigrade?

What is the absolute zero of temperature?

State the law of Boyle and Marriotte?

At what temperature and pressure has H been liquefied?

Oxygen.

Where and in what quantities is O found?

How can we get O from the air?

Upon what reaction is this method based?

From what natural substance is it often prepared?

Write the equation representing the decomposition?

What was the method used by Priestley?

What is the best method for laboratory use?

What are the properties of O?

What happens to it when it is much cooled down and compressed?

How does oxygen behave towards other substances at the ordinary temperature?

Does O act on anything at the ordinary temperature?

Give examples?

How does it act at higher temperature?

By what is this action generally accompanied?

What is this process called?

When a substance burns in O is the oxygen lost?

What becomes of it?

Do substances gain or lose in weight?

How does the weight of the O used up compare with the gain in weight of the substance burned?

State the difference between incandesence and combustion?

In what does burning in O consist?

Is burning in the air, the same chemical act as burning in O?

Give an example of a substance that will not burn in the air and will burn in O?

How was this shown?

What is meant by the kindling temperature?

Explain why a stick of wood burns gradually and not all at once?

Explain the connection between the heat and light produced, and the combustion of a substance?

What is meant by the expression, chemical energy?

Do combustible substances possess chemical energy?

Show how a combustible substance can do work?

Write equations showing the reaction of O with S, with C, with Fe, and with P?

Give the physical appearance of the several products?

What difference exists between burning a substance in the air and in O?

How can this be proved?

To what is the act of burning in the air due?

Why do not all substances burn as readily in the air as in O?

What was the phlogiston theory?

Describe the experiments of Lavoisier, upon the phenomenon of combustion?

What is meant by the term combustion, in its broadest sense?

How is this ordinarily restricted?

Define an incombustible, and a combustible substance?

Do substances combine with O without evolution of light? Give examples?

Where do we find the most important illustration of slow oxidation?

What is the chief difference between combustion, as we ordinarly understand it, and slow oxidation?

What difference is there between the quantity of heat given off when a substance burns, and when it undergoes slow oxidation?

What is the source of power in the steam engine?

23

How can H_2O and CO_2 be made to do work by combining with O?

On what is all plant life dependent?

Give an example of chemical energy stored up by the aid of the sun's heat?

What name is given to the compounds of O?

How are they divided?

Define an anhydrid, and give examples?

What is a basic oxid? Saline, neutral or indifferent oxids?

Give the analytical characters of O?

Ozone.

What change takes place in O when electric sparks are passed through it?

In what other ways can this change be brought about?

How can ozone be converted into O?

When O is changed to ozone, and ozone to O, is there any change in weight?

How many compounds of H and O are known?

What is the molecular weight of water?

How long was water considered an elementary substance?

Who first proved its compound nature?

When and by whom was its composition proven by synthesis?

What proportion of the human body is water?

How may the formation of water from O and H be brought about?

What results when two volumes of H and one volume of O are confined in a vessel and brought in contact with a flame or electric spark?

What does this show?

What is true in regard to the amount of heat produced in this chemical change?

What is meant by "Unit of heat"?

How many units of heat are set free in the combustion of one gram of H? (34.462)

How many are set free from a gram of charcoal burning?

Is there any difference in the heat developed from the burning of H in ordinary air and pure O?

Can a mixture of H and O be made to burn quietly?

Describe the compound blow pipe?

Describe the flame?

How is the calcium or oxy-hydrogen light produced?

What is the effect whenever a current of electricity is passed through a liquid capable of conducting it?

What is this method of decomposition called?

Is pure water a conductor of electricity?

How is it made capable of electrolytic analysis?

How are the two gases identified?

In the union of gases by volume, is the volume of the product the same as that of the original mixture?

What is the volume of the union of H_2 and O in the form of steam?

How near the truth is the common expression that "a cubic inch of water yields a cubic foot of steam"?

Which would require the greater quantity of heat to raise the temperature a single degree, one pound of water, or 30 pounds of quicksilver?

State the behaviour of water on cooling?

State some of the beneficial effects of expansion in the formation of ice?

What can be said of the solvent powers of water?

What results from the union of H_2O with the oxids?

Define the water of crystallization, and the water of constitution?

How is the water of crystallization indicated in chemical formulae?

What is represented by this formula, $MgSO_4$ $H_2O + 6Ag$?

What waters are best for drinking purposes?

In forming an opinion relative to the healthfulness of a drinking water, what particulars should be ascertained?

For sanitary purposes, what should be included in a water analysis?

What precaution should be taken in collecting the sample for analysis?

How would the physical properties, such as color, odor, taste, and transparency be best noted?

How is the presence of lime detected?

State the tests for chlorids, ammonia, organic matter?

What conditions in water are favorable to the solution of lead?

What natural water is most liable to contamination with lead?

How can the power of water for dissolving lead be determined?

Are ice water and snow water pure?

How can the sediment of water be obtained for microscopic examination?

State the different methods of purifying drinking water?

Describe the action of alum upon water?

How much alum to the gallon should be used?

Describe the methods of purification by the use of spongy iron, and coke?

Hydrogen Dioxid.

By what other name is this compound known?

How is it artificially prepared?

Write the equation, representing the reaction of carbon dioxid with barium dioxid suspended in water?

How is the aqueous solution of H_2O_2 concentrated?

What is the sp. gr. of the pure dioxid?

For what is H_2O_2 chiefly characterized?

For what manufacturing purposes is it now extensively used?

What distinguishes this compound from all other oxidizing agents?

What takes place when H_2O_2 comes in contact with silver oxid?

How are these changes now generally accounted for?

Describe the reaction of H_2O_2 with chromic acid?

By the laws of thermo-chemistry, what does every change tend to produce?

For what is this compound used in medicine and surgery?

How is it prepared industrially?

What are its physical properties?

What is its action on animal tissue?

At what temperature does it remain liquid?

At what temperature will it decompose rapidly?

What action has it upon arsenic, sulphids, and sulphur dioxid?

What substances decompose H_2O_2?

Is H_2O_2 found in atmospheric air?

28

The Acidulous Elements.

What are the characteristics of elements of this class?

The Chlorin Group.

What are the characteristics of this group?

By what term are they known?

What is meant by " Halogen"?

What interesting points are shown by placing the dyad salt-formers beside the monad salt-formers, with the atomic weights appended?

Has chlorin a strong affinity for O?

What is necessary to induce chlorin to combine with O?

Fluorin.

Give the symbol and atomic weight of the element fluorin?

With what is this element usually combined?

What is cryolite?

What per cent. of F does it contain?

What is the percentage of F in calcium fluoid?

How is this compound recognized?

What name did the Alchemists give to the universal solvent?

Does free F exist in nature?

How has the isolation of this element been effected?

What substances ignite spontaneously in contact with F?

With what elements does F form compounds?

Hydrogen Fluorid.

What is its symbol and atomic weight?

How is this gas obtained?

Describe HF?

Describe the method of determining the presence of F in a compound?

What compound is formed by combination of F with the silicon of the glass?

Chlorin.

Give its symbol and atomic weight?

From what is the name derived?

In what form is this element found in the mineral world?

How much Cl is contained in one pound of NaCl?

To what is this amount equivalent?

To what useful arts does NaCl furnish raw material?

Of what chemical products is it the source?

What was the "spirit of salt" of the alchemists?

How was it obtained?

What name was given to the saline mass left in the retort?

What was infered from this experiment relative to the composition of common salt?

In accordance with this view, what name was given to NaCl up to the year 1810?

Who proved that it was composed of two elementary substances?

What was also shown by the experiment of Davy?

How can Cl be prepared?

What are the physical properties of Cl?

State some of its chemical properties?

What is one of the most noticeable properties of Cl?

To what is this due?

To what are its antiseptic properties due?

What is the effect of passing NH_3 into a flask containing Cl?

31

What compound is formed when Cl acts upon ammonium chlorid?

Show by an equation that Cl is capable of removing H from a compound and taking its place. atom for atom?

From what was Cl first obtained?

What construction did Scheele put upon the result of this experiment?

What name did Scheele give to Cl?

What name was given it by Berthollet ten years afterwards?

For what is Cl extensively employed?

How is bleaching powder prepared?

Describe the latest method?

What is Labarraque's solution?

How does it act as a disinfectant?

What are the oxygen compounds of Cl?

How many oxacids of Cl are known?

How may the presence of chlorids be determined?

Acidum Hydrchloricum, (U. S. P.)

Give formula and molecular weight of Hydrogen Chlorid?

How can the relations between the volumes of H and Cl which combine with each other and the volume of the product formed be determined?

What method is used in the manufacture of HCl?

Complete the equation $H_2SO_4 + 2NaCl =$, and explain the reaction?

State the physical properties of HCl?

What are its chemical properties?

In what form is this acid usually employed in the arts and in pharmacy?

Give the sp. gr. and boiling point of a 20 per cent. solution?

What is the sp. gr. and percentage of commercial HCl?

What is the sp. gr. and percentage of the Acidum hydrochloricum, U. S. P.?

What is that of the dilute acid?

What is the difference in the percentage of the dilute acid of the U. S. and Br. Pharmacopoeias?

How is HCl classed?

What is Acidum nitrohydrochloricum, U. S. P.?

What other name is given to this compound?

What impurities are usually present in commercial HCl?

How may these impurities be identified?

If the acid is to be used for toxicological analysis, how should it be treated?

How are the chlorids formed?

What metallic chlorids are insoluble?

Give the analytical characters of HCl?

Toxicology.

What is a poison?

What is a corrosive?

When do corrosives act most energetically?

Has the degree of concentration in which the true poisons are taken any influence upon their action?

Under the above definitions how would the action of the strong mineral acids be classed?

How are their injurious results produced?

Give the symptoms of corrosion by the mineral acids?

What should be the object of the treatment?

What is the best agent for this purpose?

What should not be given?

Should the stomach-pump be used?

Bromin.

Give symbol and atomic weight?

How was this element discovered?

Give the process of its extraction from "bittern"?

In what respect is Br different from any other element?

What are its physical characteristics?

Give its sp. gr. and boiling point?

At what temperature does it become solid?

What weight of water is required to dissolve it?

Where was it first extensively used as a disinfectant, and by whom?

What compounds of Br are used in medicine?

What impurities are often found in commercial Br?

Hydrogen Bromid.

Give formula and molecular weight of this compound?

What are its properties?

How is HBr usually prepared?

What do the bromids closely resemble?

State the analytical characters of HBr?

How many oxids of Br are known?

Name the oxyacids of Br with formulæ and molecular weight?

How is HBrO obtained?

Write the equation showing the reaction?

What are some of the properties of HBrO?

How is $HBrO_3$ obtained?

To what does it correspond in properties and composition?

35

Describe $HBrO_4$?

What is noticeable about the stability of these compounds?

How is the potassium salt of bromic acid formed?

How is the potassium salt of hypobromous acid formed?

Iodin.

Give the symbol and atomic weight of Iodin?

From what is this element obtained?

Describe the process of its extraction from kelp?

How is I purified?

What is " Iodid of cyanogen"?

What is the sp. gr. of pure I?

At what temperature does it boil?

What is the test for I?

What effect has heat upon this test?

What takes place when starch mucilage is added to a solution of KI?

What will you add to the solution to obtain the iodin reaction?

How may the solubility of I in H_2O be increased?

What is Lugol's solution?

What effect has chloroform on an aqueous solution of I?

How much I will a ton of kelp yield?

What are its physical properties?

What its chemical properties?

'

Toxicology.

What is the action of I when taken internally?

How is it eliminated from the body?

What should be the treatment in case of poisoning by I?

Hydriodic Acid or Hydrogen Iodid.

Give symbol and molecular weight of this compound?

How may this substance be prepared?

Describe its physical properties?

What does the analysis of this gas teach us regarding the composition of HI?

Oxids and Oxy-Acids of Iodin.

Name the important oxy-acids and oxids of I?

Iodic Acid, or Hydrogen Iodate.

To what does this acid correspond?

Give symbol and molecular weight?

How may this acid be obtained?

Periodic Acid or Hydrogen Periodate.

How can this acid be obtained, and what are its physical characteristics?

What effect has heat upon this compound?

What does the potassium salt of this acid resemble?

Iodids.

How are the Iodids formed?

Name the soluble and insoluble metallic Iodids?

What one is slightly soluble?

What effect has Cl on the Iodids?

Give their analytical characters?

In how many proportions does I and Cl combine with each other?

38

Iodin Monochlorid, or Protochlorid.

Give the formula, and describe this compound?

How is it formed?

Iodin Trichlorid, or Perchlorid.

What are the physical characteristics of this compound?

How is it formed, and for what purpose is it used?

What is its formula and molecular weight?

Iodin and Nitrogen.

What kind of compounds result from the union of I and N?

How is the Iodid of Nitrogen prepared?

Write an equation showing the reaction?

Sulphur Group.

What elements constitute the sulphur group?

What are the characteristics of this group?

Give the symbol and atomic weight of the element S?

Where is it found in nature?

What is the chief source of the sulphur of commerce?

Name some of the principal compounds in which sulphur occurs in nature?

How is crude brimstone refined?

What is the difference between " flowers of sulphur" and " roll sulphur"?

What changes does it undergo when it is heated?

In what form does S occur in plants?

What proportion of S in vegetable albumen?

In what form does S occur in animals?

In what animal compounds is it a constituent?

Describe the process by which sulphur is extracted from its ores?

In how many forms may S be obtained?

Describe the several forms?

How may the second variety be produced?

How does this variety act towards carbon disulphid?

Why is S called a dimorphous element?

How is the third variety produced?

What effect has CS_2 upon it?

What is the difference in sp. gr. of the second and third variety?

How do you explain the fact that S at 500° has a vapor density of 96 but at 1000° it becomes 32?

At what temperature does S take fire in the air?

What are sulphids?

When does S act as a dyad?

How is S recognized in the free state?

What are the tests for S in combination?

For what purposes is S employed?

Hydrogen Sulphid.

What is the hydrogen compound of S and how may it be prepared?

Where is this compound found in nature?

Describe H_2S?

What action has this gas on the lead compounds?

What use can you make of this test in cases of defective plumbing, or exhalations from cesspools, etc.?

What does H_2S form with metals?

What odor is given off when a sulphid is heated?

Does H_2S occur in the body?

Where and how does it form?

How does H_2S act as a poison?

What is the treatment for poisoning with H_2S?

To what are its toxic powers due?

In what form does H_2S generally produce deleterious effects?

Oxids of Sulphur.

Give symbol and molecular weight of sulphur dioxid?

State the different methods of its preparation?

For what purpose is the second method usually employed?

Why is there so small a percentage of SO_2 in the air of manufacturing towns?

Why is the fifth method of interest to medical students and pharmacists?

Describe SO_2?

State the chemical characteristics of SO_2?

How can it be shown that this gas is easily absorbed by water?

What proportion of its bulk is absorbed by H_2O?

Is SO_2 a supporter of combustion?

What are some of the principal uses of SO_2?

What articles are usually bleached by this gas?

How is the bleaching effected?

What is meant by the antiseptic or antizymotic properties of SO_2?

How is this turned to account in the manufacture of wine and beer?

What action has SO_2 on the sulphites and on solutions of silver and gold?

What is the effect of heating Fe, Sn, Pb, and Zn, in SO_2?

Sulphites.

What general resemblance is there between the sulphites and carbonates?

For what is sodium sulphite used?

How is it prepared?

Give the analytical characters of the sulphites?

Sulphur Trioxid or Anhydrid.

Give formula and molecular weight?

How is it prepared?

What are its properties?

Hydrosulphurous Acid.

Give formula and molecular weight?

How is this compound obtained?

What are its properties?

Sulphuric Acid.

Give the formula, molecular weight and sp. gr. of this acid?

What is its boiling temperature?

How is it prepared?

How is the weak chamber acid concentrated?

Describe the commercial oil of vitriol?

What is the C. P. Acidum sulphuricum of the U. S. P?

Describe the Glacial sulphuric acid?

How much H_2SO_4 is contained in the dilute acid of the U. S. P.?

How does the dilute acid of the Br. P. differ from the U. S.?

Give the sp. gr. of each?

Describe H_2SO_4?

State some of its chemical properties?

What are some of the impurities and how are they identified?

When H_2SO_4 is used for toxicological analysis, how is its purity to be tested?

Give the analytical characters of H_2SO_4?

44

Toxicology.

Is H_2SO_4 a corrosive or a true poison?

How may death be caused by the concentrated acid?

What is the treatment?

Thiosulphuric Acid or Hydrogen Thiosulphate.

Give the symbol of this compound?

What is the formula of a metallic thiosulphate?

For what is sodium thiosulphate used?

How is this salt prepared?

What is Nordhausen or fuming sulphuric acid?

Selenium.

Give symbol and atomic weight of this element?

In combination with what elements is it usually found?

What can be said of its compounds and allotropic forms?

Tellurium.

Give symbol and atomic weight?

What can be said of its compounds?

Nitrogen Group.

What are the characteristics of the nitrogen group?

What elements comprise this group?

Give the symbol and atomic weight of nitrogen?

By whom discovered and when?

What was the first name given to this gas?

Why was the name *Nitrogene* given to it later?

Where is N found free and in combination?

What is the most convenient method to prepare N?

Show by an equation what takes place when Cl is passed into a solution of H_2O and NH_3?

Why should this experiment be made with caution?

Give the physical properties of N?

Does N combine readily with the other elements?

Is it a supporter of combustion and respiration?

Why do animals die in an atmosphere of this gas?

What purpose does it serve as a constituent of the air?

What would result if the air consisted only of O?

What is the composition of the atmosphere, by volume and by weight?

Is the air a chemical compound or a mechanical mixture?

State the evidence showing why the last answer is correct?

Is the quantity of O in the air decreasing?

What is involved in the process of plant life?

Explain how in the process of vegetable growth we have a constant source of oxygen supply?

What are some of the offices of Nitrogen in the processes of nature?

Have plants the power to take up from the air a part of the N they need?

Is it possible that other agents than plants are concerned in the absorption of N from the air?

What is pure air?

What is the most common cause of its contamination?

What is involved in the breathing process?

In what way does reducing the quantity of O produce evil effects?

To what is the ill effect of breathing the air of a badly ventilated room due?

How do these organic matters act on the system?

Name a source of supply of pure air in most buildings, independent of architect's plans?

What results from breathing air badly contaminated by the decomposition of animal and vegetable matter?

What is meant by the term malaria?

What is the real source of trouble in breathing this *bad* air?

Ammonia.

Give formula and molecular weight?

How is NH_3 obtained ?

What proportion of NH_3 exists in the air?

By what means do plants derive their chief supply of N from the atmosphere?

How is NH_3 formed from atmospheric N?

What is the chief source of Liquor Ammonia?

State the process of purification?

What are the principal properties of NH_3?

How many volumes of NH_3 are absorbed by one volume of H_2O?

Is this a case of chemical combination or solution?

Why was the name " hartshorn " given to NH_3?

Write an equation showing the result of the union of Ammonia, calcium monoxid and oxygen, and state the principle it illustrates?

What is meant by nitrification?

State what is at present known relative to the nitrifying ferment?

What results when ammonia-water is mixed with nitric acid?

Why is NH_4 called *ammonium?*

Nitrogen Monoxid.

Give formula and molecular weight of this compound?

How is this gas prepared?

Describe nitrogen monoxid?

How can it be distinguished from O?

What is the result of a mixture of equal volumes of N_2O and H ignited by the electric spark?

What is the special property of N_2O?

Nitrogen Dioxid.

What is the molecular weight and formula of this compound?

How is NO prepared?

49

Complete the equation, $8HNO_3 + Cu_3^- = ?$

Describe the properties of NO?

Nitrogen Trioxid.

What other name has this compound?

Give molecular weight and formula?

How is N_2O_3 obtained?·

Nitrogen Tetroxid.

What other names has this compound?

What is its formula and molecular weight?

How is NO_2 prepared?

Complete the equation $Pb\ (NO_3)_2 =$
and explain the reaction upon the application of
heat to the lead nitrate?

What are some of its properties?

To what is the oxidizing powers of the red
fuming nitric acid due?

What is the so-called *nitrous acid* of commerce?

How is it prepared?

Complete the equation, $2NO_2 + H_2O = ?$

Toxicology.

What caution should be observed in those processes in which HNO_3 is decomposed?

What should be done when this gas is discharged into the air in large quantities?

Which is the most dangerous to health, NO_2 or Cl?

What are the symptoms of poisoning with NO_2?

What does the autopsy show?

How may accidents from inhaling NO_2 be prevented?

What should be done when nitric acid is spilled in the laboratory?

Nitrogen Pentoxid.

Give formula and molecular weight?

How is this compound prepared?

What are some of the properties of N_2O_5?

Nitrogen Acids.

How many nitrogen acids are known?

Give the names and formulæ of each?

Describe Hyponitrous acid?

Describe nitrous acid?

Nitric Acid and Other Nitrates.

Give molecular weight of HNO_3?

By whom, and in what year, was its composition determined synthetically?

What is meant by the *nitric radical*?

To what is the production of nitrates due?

Where are nitrates commonly found?

What is meant by "prismatic nitre"?

How is it obtained?

For what uses is Potassium nitrate employed?

Sodium Nitrate.   ·

What other name has this compound?

Where is it found?

Why is this product unfit for making gunpowder?

How is Potassium nitrate made artificially?

How is HNO_3 prepared?

52

Complete the equation $KNO_3 + H_2SO_4 =$
and explain the reaction?

State the properties of HNO_3?

How may pure HNO_3 in its most concentrated
form be obtained?

State the action of dilute and concentrated
HNO_3 upon many metals and organic substances?

For what are the nitrates remarkable?

What impurities are often present?

How may they be detected?

What is the precipitant for HNO_3?

What is the best method for the detection of
HNO_3?

What uses has HNO_3 in the arts?

What is Aqua regia?

What is the Nitro-hydrochloric acid of the
U. S. P.?

Toxicology.

To what is the poisonous action of the nitrates
due?

What is the action of HNO_3 as a poison?

What effect has the concentrated acid on animal
tissue?

What is its action when taken internally?

What symptoms are developed?

What treatment should be followed?

What is the sp. gr. and percentage of the strongest acid met with in commerce?

What is the composition of the officinal HNO_3?

What effect has heat on a weaker or stronger acid?

Compounds of Nitrogen with the Halogens. Nitrogen Chlorid.

Give the formula and molecular weight of this compound?

What are its properties, and how is NCl_3 produced?

How many gram-degrees of heat per equivalent is absorbed in its formation?   (38478)

How does this fact explain the violent energy of this compound?

To what is its instability attributable?

How is its explosion brought about?

At what temperature has NCl_3 been distilled without explosion?

Nitrogen Bromid.

Give formula and molecular weight?

How is NBr_3 obtained?

What are its properties?

54

Nitrogen Iodid.

Give molecular weight and formula of this compound?

How is NI_3 prepared?

What are the properties of NI_3?

Phosphorus.

Give symbol and atomic weight of this element?

How does P occur in nature?

Why are the phosphates the most important compounds of P?

How are the phosphates which are taken into the body given off?

In what, and when, was P first discovered?

How is P manufactured?

How is the crude P purified?

Describe the ordinary P?

What results when a solution of P in carbon disulphid is added to a solution of copper sulphate?

What is amorphous P?

How does it differ from ordinary P?

How is it converted into ordinary P?

For what is P used in the arts and medicine?

55

Toxicology.

How does amorphous P differ from the other forms of P in its toxic qualities?

How should burns from P be treated?

What care should be taken in handling P?

What can be said of the poisonous action of yellow P?

Give the symptoms of P poisoning?

What chemical antidote should be given?

What should be the treatment?

What complication is often met with in cases of P poisoning?

Analysis.

How soon after a death, supposed to be caused by P, should the investigation be made?

Why is this necessary?

Describe the method of analysis?

What is the Lucifer disease?

What is the culminating manifestation of this disease?

How may the frequency of this disease be diminished?

How can perfect prevention be obtained?

Phosphids of Hydrogen.

Name the compounds of P and H?

Give the formulæ of each?

Which of these compounds is the most important?

How is PH_3 produced?

What are its properties?

In a case of poisoning by PH_3 what is the condition of the blood after death?

What are the properties of P_2H_4?

Describe P_4H_2?

Oxids of Phosphorus.

How many oxids of P are known?

Give formulæ of each?

How is P_2O_3 obtained?

What are its physical properties?

How is P_2O_5 produced?

Which of these oxids is the acid-forming oxid?

What are its properties?

Phosphorous Acids.

How many oxyacids of P are known?

Name them and give formulæ?

Describe the properties of H_3PO_2?

What is the molecular weight of H_3PO_2?

How is H_3PO_3 formed?

What are its physical properties?

How is H_3PO_4 prepared?

Describe H_3PO_4?

Into what is it decomposed by heat?

Give the analytical characters of the ortho-phosphates?

How is $H_4P_2O_7$ formed?

Give formula for Glacial phosphoric acid?

How is it formed?

What other name has it?

What does the word "Meta" indicate?

What are the properties of HPO_3?

What acid does it resemble in constitution and basicity?

———————

Compounds of Phosphorus with the Halogens?

———

How many compounds of P with Cl?

Name them and give formulae?

How is PCl_3 obtained?

What are its properties?

How is PCl₅ formed?

What are its properties?

Complete the equation $PCl_5 + H_2O =$ and describe the product?

Name the compounds of P with Br, I, and Fl?

Arsenic.

Give symbol and molecular weight of Arsenic?

Why is this element often classed with the metals?

What is its mode of occurrence in nature?

What is the composition of arsenical pyrites?

What other arsenic compounds deserve mention?

What is the appearance of arsenic?

What effect has heat upon As?

What is the characteristic odor of As?

How does As combine with most elements?

Is arsenic, as an element, poisonous?

What change takes place when As is boiled with HNO_3?

To the compounds of what element, are some of the compounds of As analogous?

Like what element does As conduct itself in nearly all its compounds?

Compounds of Arsenic and Hydrogen.

How many compounds of As and H are known?

Give molecular weight of AsH_3?

To what is this compound analogous?

How is it prepared?

What are its chemical properties?

When ignited in the air what are the products of the combustion?

If burned without free access of air what results?

What substances decompose AsH_3?

What results when AsH_3 is passed into a solution of a metallic salt?

Oxids of Arsenic.

How many compounds does As form with O?

To what compounds of P do these correspond?

Give formula and molecular weight of arsenious oxid?

What other name is given As_2O_3?

What uses are made of As_2O_3 in the arts?

How is As_2O_3 manufactured?

What is the sp. gr. of this compound?

What is its appearance under the microscope?

What effect has heat on As_2O_3?

When As_2O_3 is sprinkled upon red hot coal what odor is developed?

To what is this due?

Is As_2O_3 easily soluble in H_2O?

How many grains of As_2O_3 would be taken up by a pint of H_2O?

If boiling water be used and allowed to remain in contact till cold, how many grains will dissolve?

Is As_2O_3 soluble in glycerin?

What takes place when As_2O_3 is dissolved in hot HCl?

How many varieties of arsenious oxid?

What is formed by As_2O_3 dissolved in solutions of the alkalies?

What takes place when HCl is added to the solution of an alkaline arsenite?

Give formula of Arsenic Pentoxid, with molecular weight?

Describe its appearance?

What effect has heat upon As_2O_5?

Complete the equation $As_2O_5 + 3H_2O = $?

To what do the arsenates correspond?

Arsenites.

What is the basicity of arsenious acid?

Which are the more stable, arsenates or arsenites?

For what is sodium arsenite used?

What is Scheele's green?

How is it prepared?

For what is it used?

What is its effect on health?

How is the presence of arsenite of copper proved in a sample of wall paper?

What is Emerald green?

For what is As_2O_3 often used by grooms?

Arsenic Acid.

Give formula for this compound?

For what is H_3AsO_4 used in the arts?

How is H_3AsO_4 prepared?

Give formula for Sodium arsenate?

What is its use in the arts?

How is it manufactured?

Which is the more powerful acid, Arsenious or Arsenic?

What action has H_2SO_4 on Arsenic acid?

Complete the equation, $H_3AsO_4 + H_2SO_4 = ?$

Give the composition of the sulphids of As?

Give molecular weight of As_2S_2?

Describe the mineral?

How is the compound used in the arts generally prepared?

For what is this compound used?

Give formula and molecular weight of Arsenic trisulphid?

What is the composition of King's yellow?

Give molecular weight of As_2O_3?

How is this sulphid obtained?

Compounds of Arsenic and the Halogens.

Give formula and molecular weight for Arsenic trifluorid?

How is it prepared?

What are its properties?

By what is it decomposed?

Give formula and molecular weight for Arsenic trichlorid?

How is it produced?

What are its properties?

Give formula and molecular weight for Arsenic triiodid?

How is this compound formed?

Describe its physical appearance?

By what is it decomposed?

63

Action of Arsenical Compounds Upon the Animal Economy.

What are the effects of elementary arsenic upon the system?

By what means does it become oxidized?

What is the most actively poisonous of the inorganic compounds of As?

What compound of As is most frequently used by criminals?

How has it been given?

If the poison has been given in quantity in what condition is it often found in the stomach?

What is the fatal dose?

Has the condition of the stomach any effect on the activity of the poison?

How many fatal cases have been reported as due to Fowler's solution?

In what way has Sodium arsenite caused death and illness?

How many deaths have been directly traced to Arsenic acid?

Are the cases of death and illness which have been put to the account of anilin dyes, due to them directly?

To what are they due?

To what is the poisoning by sulphids of arsenic due?

What can be said of the use of the Arsenical greens, and of their administration with murderous intent?

How is the poison from room paper, hangings, etc., disseminated in the atmosphere?

What should be the treatment for acute arsenical poisoning?

What is the object in administering a chemical antidote?

What is the chemical antidote and how is it prepared?

How is it to be given and in what doses?

Precautions to be Taken by the Physician in Cases of Suspected Poisoning.

What is the first duty of the physician who suspects poisoning, during the life of the patient?

What should he examine and in what way?

If the case terminate fatally what should he do?

At the post mortem investigation, who should be present with the physician?

Why is this necessary?

In cases where the chemist cannot be present, what must be remembered?

Would the finding of poison in the stomach be sufficient to procure conviction?

What parts should be sent to the chemist and how should they be sealed?

What precautions should be taken about vessels, corks, caps and seals?

If the physician fails to observe these precautions, what has he done?

Give the analytical characters of the arsenical compounds?

What is Reinsch test for As?

Describe it?

What are the advantages of this test?

Why should not this method be used after death?

What precautions are to be taken when making this test?

On what reaction is Marsh's test based?

Describe Marsh's test for As?

How can the stains from As and Sb on the porcelain be distinguished from each other?

How many arsenical preparations has the U. S. P.?

Name them and give the composition of each?

Antimony.

Give atomic weight and formula?

How does Sb occur in nature?

From what mineral is the antimony of commerce obtained?

Describe the metal Sb?

What is its melting temperature?

For what is Sb used in the arts?

What quality does Sb communicate to the compound with which it is alloyed?

How many compounds of Sb with H?

Describe H_3Sb?

How many compounds of Sb and O?

Name them and give their formulæ?

How many compounds of Sb and Cl?

Describe $SbCl_3$?

How is $SbCl_5$ formed?

Give formula and molecular weight of antimony trisulphid?

How may it be formed?

What is Kermes mineral?

Describe Sb_2S_5?

Give formula for "Tartar Emetic"?

How is it obtained?

Complete the equation $2KHC_4H_4O_6 + Sb_2O_3 =$?

What is the wine of antimony of the U. S. P.?

Action of the Antimony Compounds on the Economy.

What compound is the most frequent cause of antimonial poisoning?

What is the smallest fatal dose of this compound?

What is the effect of small and repeated doses?

How can the physician satisfy himself, if he suspects antimonial poisoning?

What should be the treatment, in a case of antimonial poisoning?

State the analytical characters of the antimonial compounds?

Boron.

Give symbol and atomic weight?

What are the characteristics of this element?

Where is B chiefly found?

In how many forms does the element B exist?

Upon what type are the compounds of B formed?

Where is Boric acid chiefly obtained?

Describe the method of manufacture?

For what is Boric acid used?

Name some of the important uses of Borax?

What takes place when Borax and strong H_2SO_4 are mixed?

How may Boric acid be detected analytically?

Carbon Group.

What elements constitute the carbon group?

What are the characteristics of this group?

Carbon.

Give atomic weight and symbol of Carbon?

Why is carbon called the central element of organic nature?

In how many forms in nature does C occur?

What is coal?

When and how were the energy of our coal-beds accumulated?

From what source came this energy?

Where does this mysterious action take place?

What is the chief article of diet in plant life?

Explain the decomposition of CO_2 into C and O?

Where does the energy expended by the sun in pulling apart the O and C atoms reappear?

How much latent energy in each pound of coal?

What reason have we to state that C is one of the most fixed solids known?

Of what elementary substances do plants and animals chiefly consist?

How may all organized beings be described?

What is meant by amorphous carbon?

What is charcoal?

If wood is heated in a closed vessel what are the products formed?

What is this process called?

When chemical compounds are heated how do the constituents tend to arrange themselves?

What is the product of the combustion of charcoal in the air or O?

What product is produced when there is a lack of O, or in a bad draught?

In what is charcoal soluble?

What is coke?

From what is it manufactured and how?

What is lamp-black?

How is it produced, and its use in the arts?

What is bone-black?

How is it made?

How is the mineral matter removed?

On what does the power of charcoal to absorb gases depend?

How may most coloring matters in liquids be removed?

What proof can you give that charcoal and graphite are chemically identical with the diamond?

How much CO_2 is obtained from the combustion of one gram of pure charcoal?

What properties have the three forms of C in common?

Why is C called a reducing agent?

For what is it extensively used in the arts?

What is meant by the term Hydro-carbons?

On what does the value of coal for the manufacture of gas depend?

By what names is the compound CH_4 known?

How is it produced in nature?

What is the miner's name for this gas?

What is Ethylene?

How does the density of C_2H_4 compare with that of CH_4 and with H?

What were the old names for CH_4 and C_2H_4?

What are the characteristics of C_2H_2?

Complete the formula, $C_2H_2 + H_2 =$?

What is produced by the union of Cu or Ag with C_2H_2?

When a candle burns with a smoky flame, to what is its peculiar odor due?

In ordinary gas, upon what does the illuminating value depend?

Describe the structure of a flame?

What is meant by the oxidizing flame?

What by the reducing flame?

Describe the method of obtaining, with the blow-pipe, the reducing and oxidizing flames?

What is the oxy-hydrogen blow-pipe?

Explain the principle of the Bunsen burner?

What is implied by incandescence?

To what is the color of flame due?

Silicon.

Give symbol and atomic weight of this element?

In how many different forms may this elementary substance be obtained?

How may pure Si be obtained?

Is Si ever found in the free state?

What compounds does Si form with H, Cl, and F?

Give formula and molecular weight for Silicic oxid?

What is its purest native form?

What does it form when heated to redness with the alkaline carbonates?

Give formula and molecular weight of Hydro-fluosilicic acid?

How is it obtained in solution?

For what is it used?

Give the members of the Vanadium Group?

Give the symbols and atomic weight of each?

Name the members of the Molybdenum Group?

Give symbols and atomic weight of each?

For what is phosphomolybdic acid used?

What use is made of osmic acid?